Sweety finds her Family

Written by Katie Godfrey
Illustrated by Julia Gootzeit

To the sweetest first graders!
Katie Godfrey

Green Frog Publishing • East Montpelier, VT

Sweety Finds Her Family

Written by Katie Godfrey

Green Frog Publishing
P.O. Box 46
East Montpelier, VT 05651

www.greenfrogpublishing.com

First Edition

Written by Katie Godfrey
Illustrated by Julia Gootzeit
Layout and Cover Design by Cecilia Bizzoco
Katie Godfrey photos by Ina Larsen
Julia Gootzeit photo by Eric Gregory
Sarai Rodríguez photo by Macarena Falcón Marrero

ISBN-10 0615886450
ISBN-13 978-0-615-88645-9

Thank you to the Wormfarm Institute for supporting a rich community of artists in the rolling hills of Wisconsin.

Sweety was raised in a wonderful family.

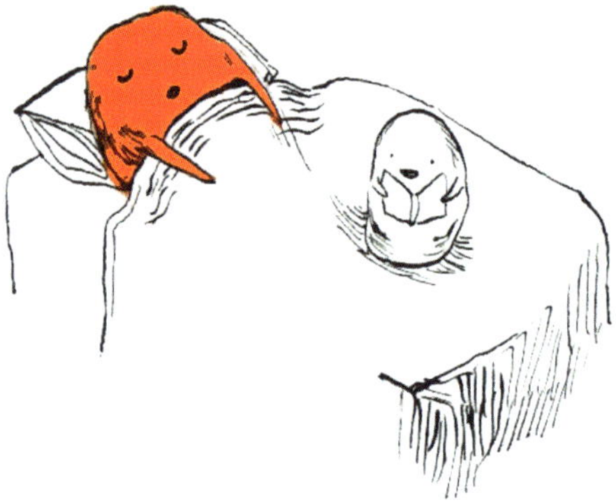

1…2…3…

Her parents read bedtime stories to her every night.

She went to the beach with her brothers and sisters.

On Saturdays the whole family played hide and seek together.

Sweety loved her family very much, still she knew that she was different. She was bigger than her brothers and sisters, in fact, Sweety was even bigger than her parents!

She was much more colorful than the rest of her family. And she liked to wake up early, before everyone else to catch the sunrise.

Doesn't 'sol' mean 'sun' in Spanish?
Si! Those Solanaceaes sure like the sun!

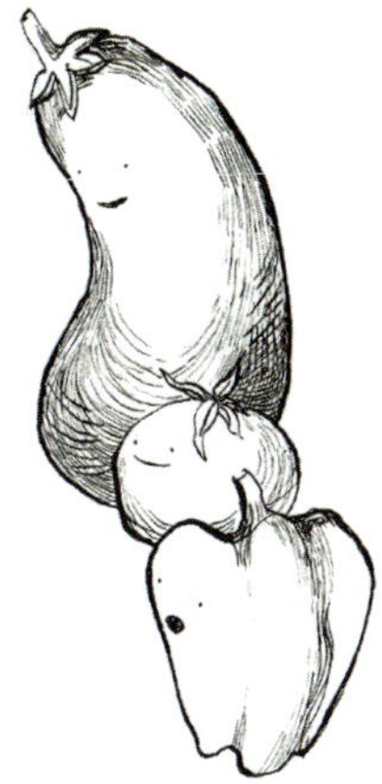

Her parents had explained a long time ago the reason why Sweety looked and felt so different.

"We found you in our garden when you were just a little girl, and we quickly realized that you're a Sweet Potato. We Potatoes belong to the **Solanaceae** (so·la·na·see·ee) family. Our cousins are the Tomatoes, Peppers and Eggplants, but no Sweet Potatoes. Our family likes hot weather and warm soil, just like you, but otherwise we're very different."

Young Sweety didn't understand how a Potato could be related to any of those other plants. Sweety might be different, but she certainly looked more like a Potato than the Tomatoes, Peppers, and especially the Eggplants, did.

Her parents explained, "Like most other plant families, when we're adults, we don't look very much alike, but when we're babies and children, we look almost exactly the same."

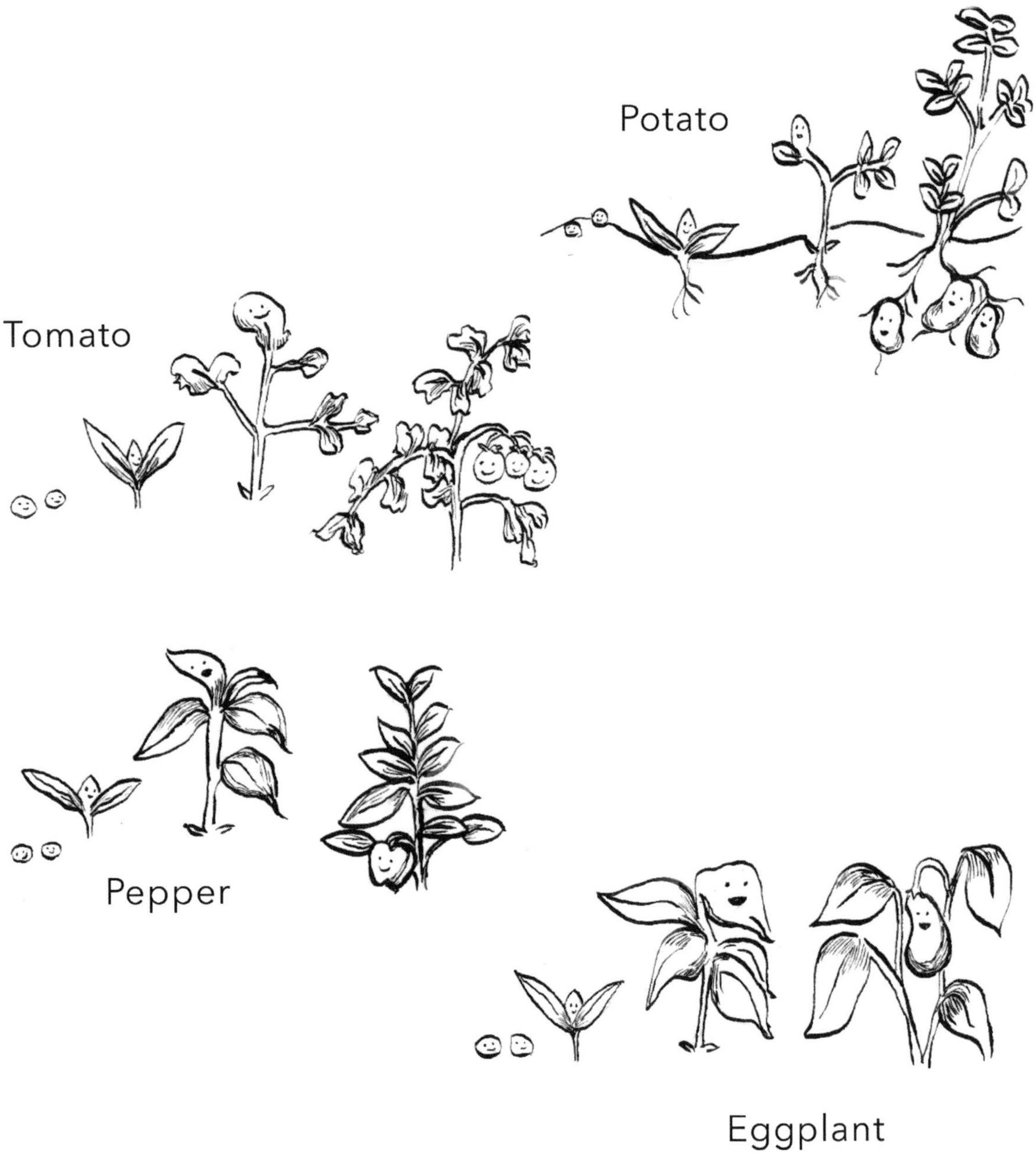
Potato
Tomato
Pepper
Eggplant

So Sweety, a Sweet Potato, grew up with the Potatoes. She was a happy girl, but always curious about her biological family.

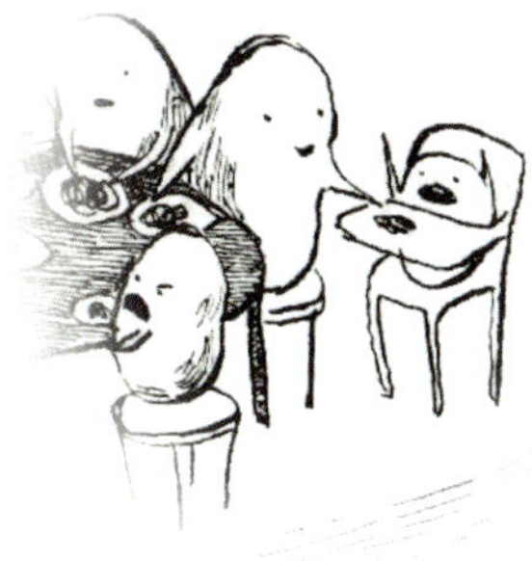

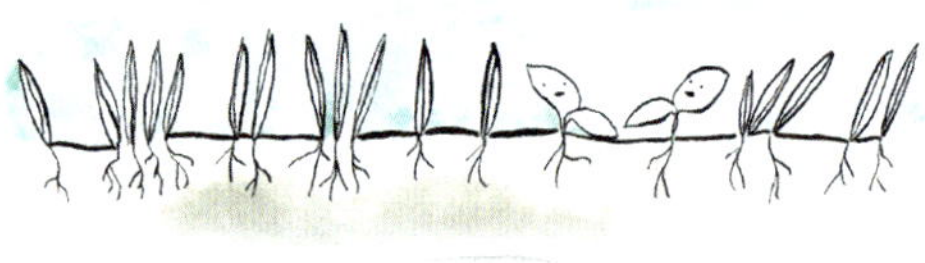

One morning, as Sweety lie awake in bed while everyone else slept, she made a decision. She was going to find her Sweet Potato family. Sweety packed a backpack and said a tearful goodbye to her parents, brothers and sisters. Then, Sweety set off on her journey.

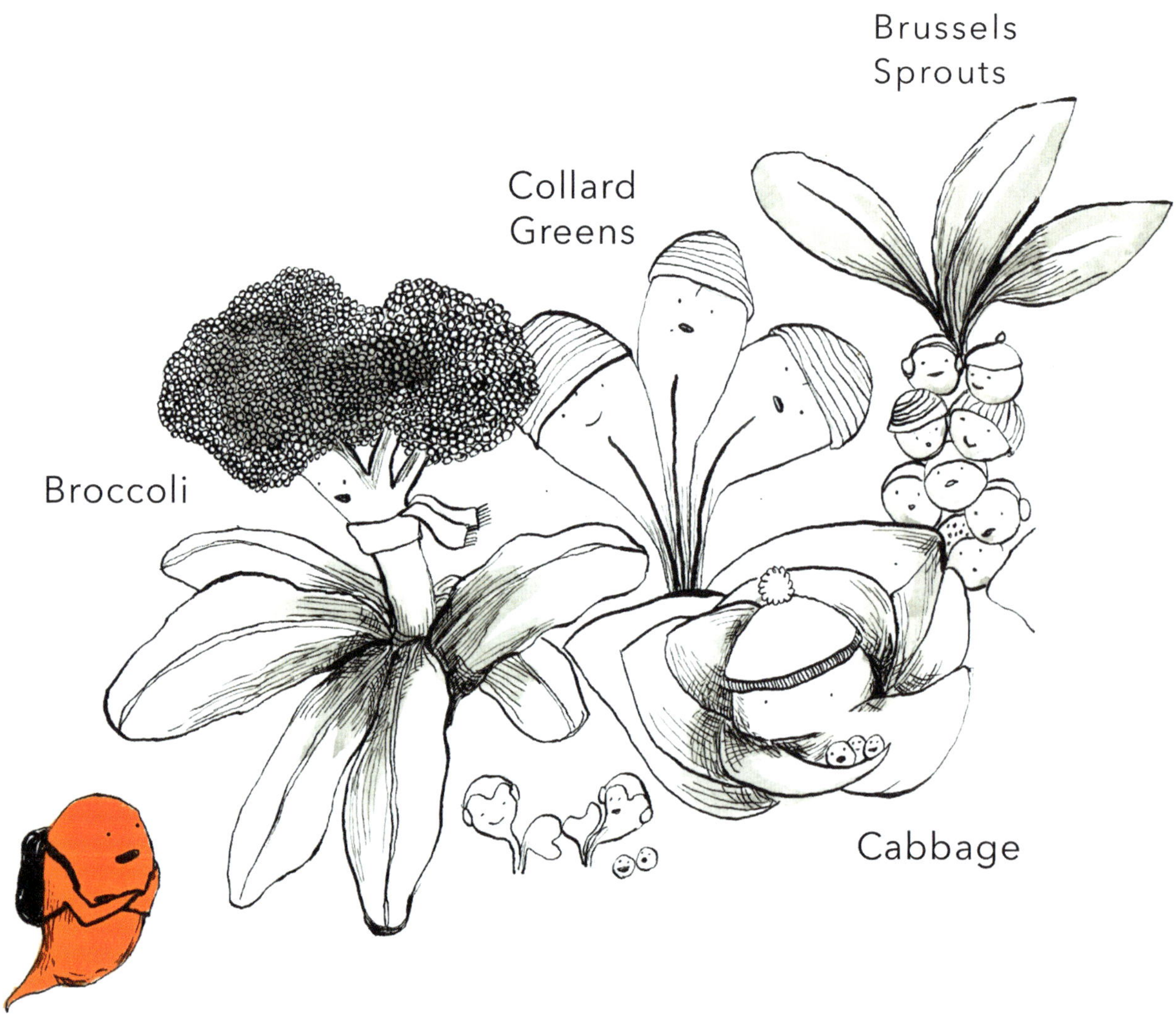
Brussels Sprouts
Collard Greens
Broccoli
Cabbage

She walked and walked until she came across her neighbors, the **Brassicas** (bras·si·cas).

"Are you my family?" asked Sweety.

No, Sweety," said Mrs. Broccoli. "The **Brassica** family has Broccolis, Cauliflowers, Cabbage, Brussels Sprouts, Rutabagas, Turnips and Radishes. We may have a large family, but no Sweet Potatoes. You wouldn't want to live with us anyway—you see, we like cold weather. Have you asked the **Allium** (al·li·um) family? They like to live underground like you!"

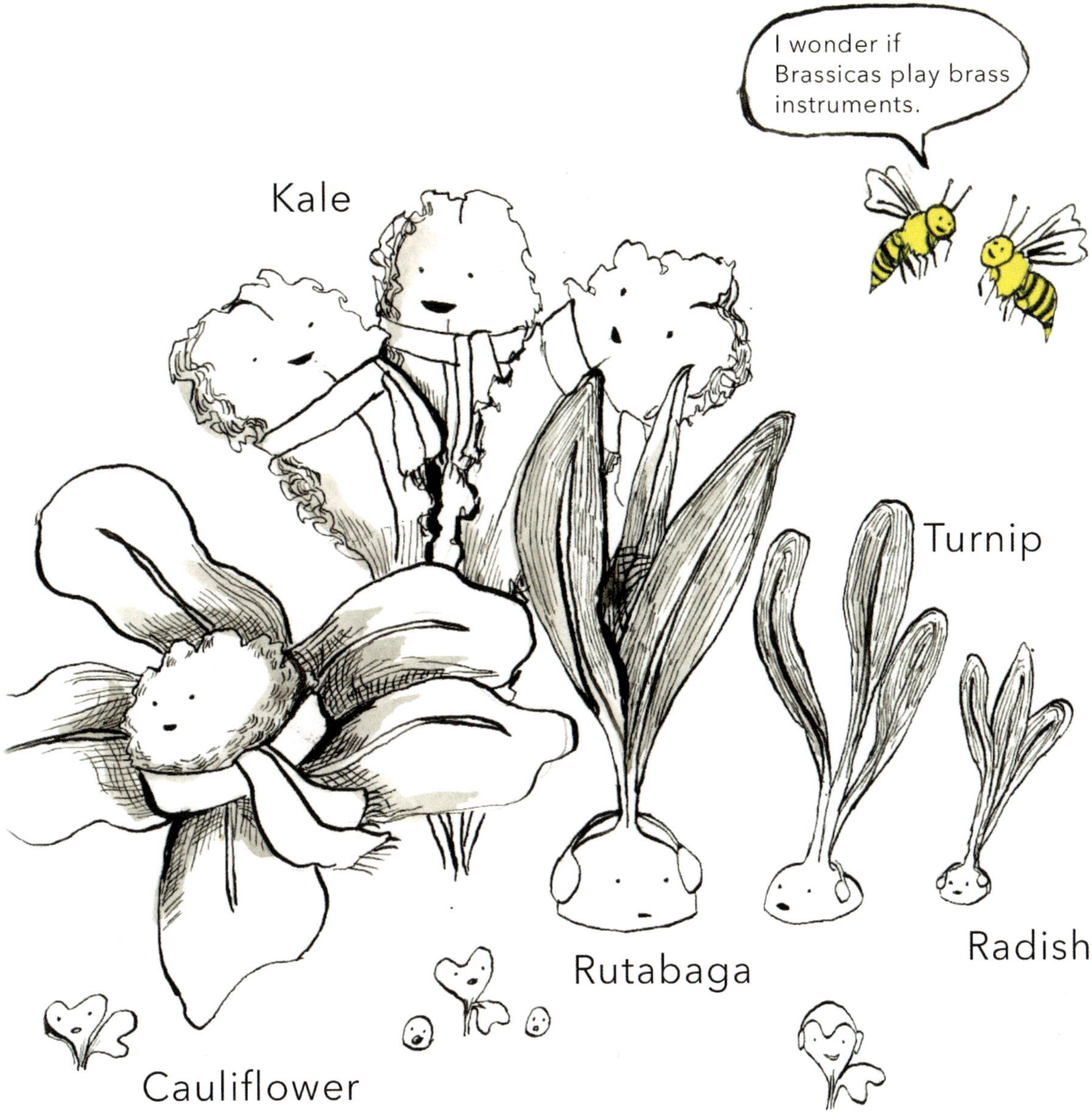
I wonder if Brassicas play brass instruments.
Kale
Turnip
Radish
Rutabaga
Cauliflower

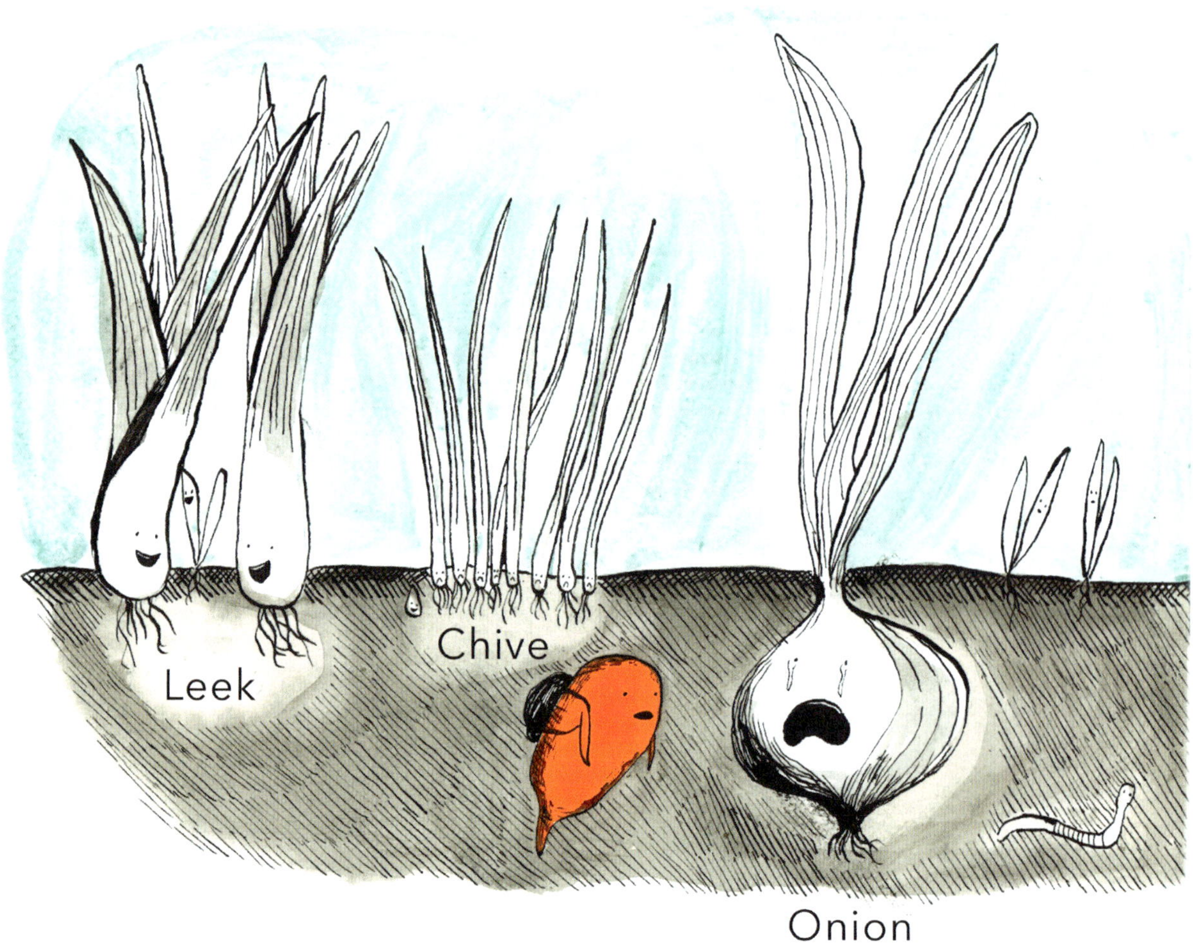
Leek
Chive
Onion

So Sweety walked and walked until she saw a family living underground.

"Are you my family?" asked Sweety.

"No, Sweety," said Mr. Onion between sobs. "We have Onions, Garlic, Leeks, Chives, Shallots and Scallions in the **Allium** family... no Sweet Potatoes though. Our leaves point up to the sky and we like to sleep underground for a long time. Plus, we're a very emotional family, with all our tears. Why don't you ask the **Cucurbit** (cu·cur·bit) family? They're big, but they're shaped a bit like you."

Alli...
Ummmm...
Garlic
Shallot
Scallion

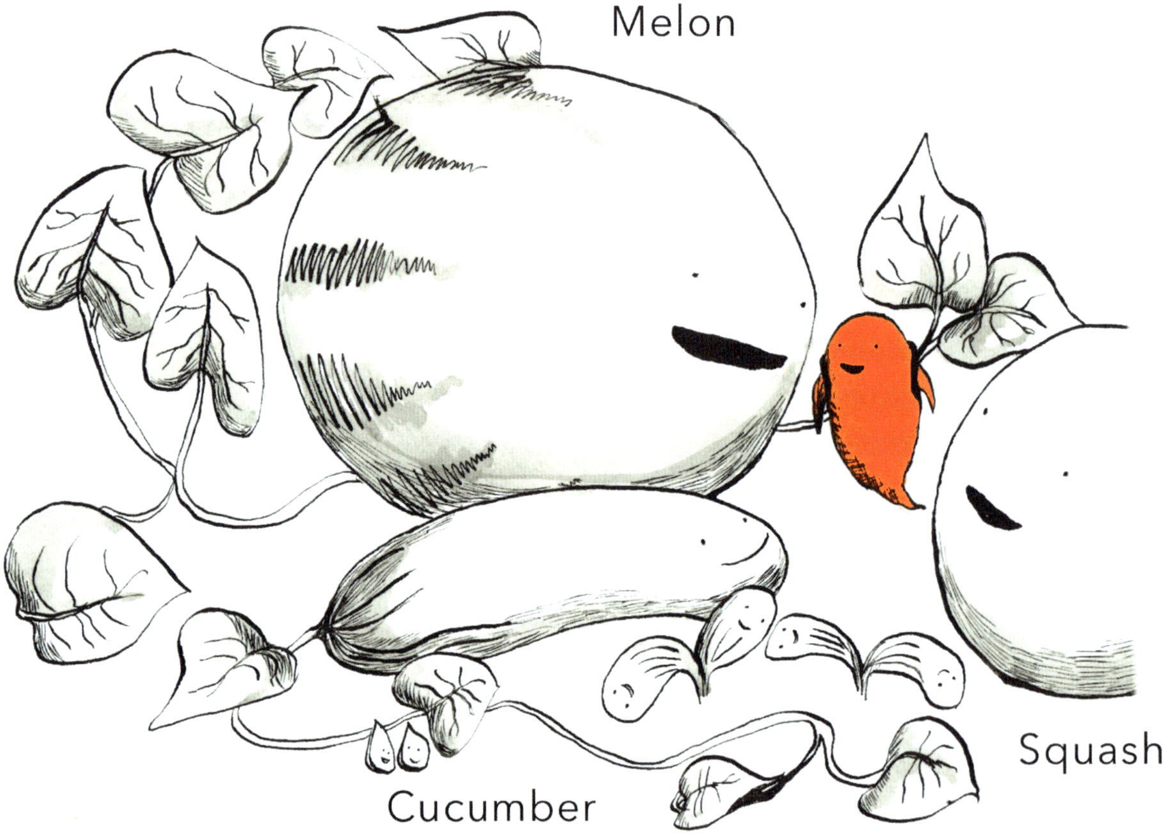
Melon
Squash
Cucumber

Sweety walked and walked until she found a family of giants.

"Are you my family?" asked Sweety.

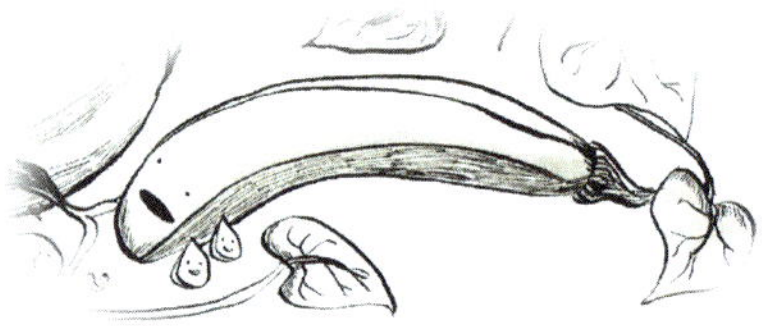

No Sweety," said Mrs. Melon. "The **Cucurbit** family includes the Melons, Cucumbers, Summer and Winter Squashes—no Sweet Potatoes here. We're much larger than you, and we all live above the ground. Why don't you ask the **Amaranth** (am·a·ranth) family? They're very healthy, just like you."

Pumpkin
Cu-cur-bit.
Sounds like cucumber!
Zucchini

Beet

Sweety walked and walked until she saw a very healthy-looking family.

"Are you my family?" asked Sweety.

No, Sweety," said Mrs. Beet. "Just the Beets, Chards and Spinach in the **Amaranth** family, but no Sweet Potatoes. Our leaves are wide and thick – they're full of healthy nutrients. Why don't you ask the **Apiaceae** (ay·pee·AY·see·ee) family? They're very pretty, like you!"

I'm a ranth!
What's a ranth?
Spinach
Swiss Chard

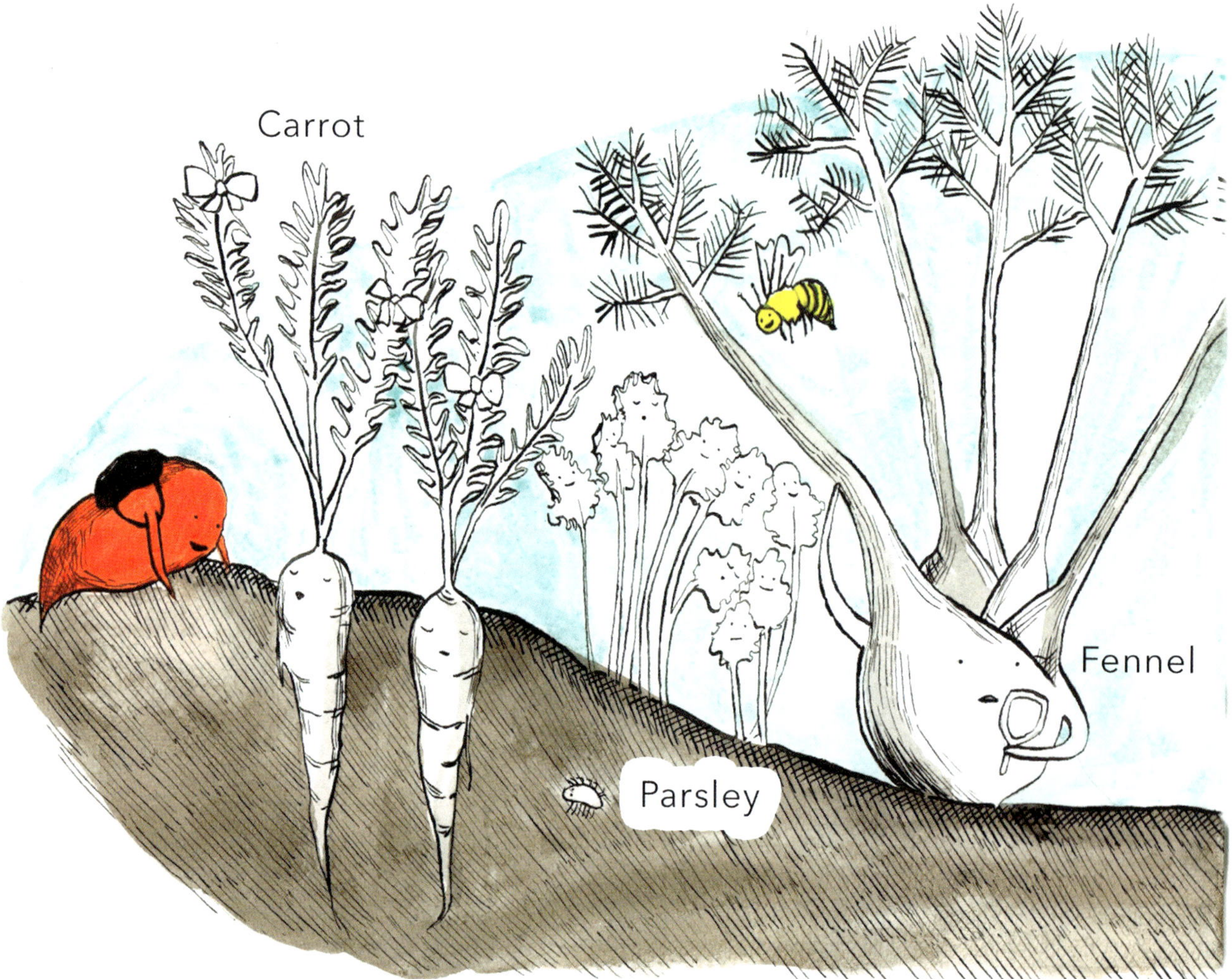
Carrot
Fennel
Parsley

So Sweety walked and walked until she found a very pretty family.

"Are you my family?" asked Sweety.

No, Sweety," said Mrs. Carrot. "Our beautiful **Apiaceae** family includes the Carrots, Parsnips, Celeries, Dills and Cilantros. We have tap roots and umbrella tops, and we look and smell good to insects that help pollinate other plants. Why don't you ask the **Asteraceae** (ass·ter·AY·see·ee) family? They're very happy, like you!

Celery
Dill
Cilantro
Apia-like apiary, the place where honeybees live.
That explains why we like them so much!
Parsnip

Sunflower
Lettuce
Marigold

So Sweety walked and walked until she found a very happy-looking family.

"Are you my family?" asked Sweety.

No, Sweety," said Mr. Lettuce. "In the **Asteraceae** family, we have Lettuce, Sunflowers, Marigolds and Daisies. Our flower petals are shaped like stars. Why don't you ask the **Fabaceae** (fab·AY·see·ee) family? They're very important and wise."

Dahlia
I stare at ya!
Asteracea!
Zinnia

Green Bean
Pea

So Sweety walked and walked until she found a family who looked very important and wise.

"Are you my family?" asked Sweety.

No, Sweety," said Mrs. Pea. "The Peas, Beans, Lentils, and even Peanuts, make up the **Fabaceae** family. Some people call us the **Legumes**. We live in pods that allow special bacteria to grow on our roots. The special bacteria takes a gas called nitrogen from the air and turns it into food for other plants."

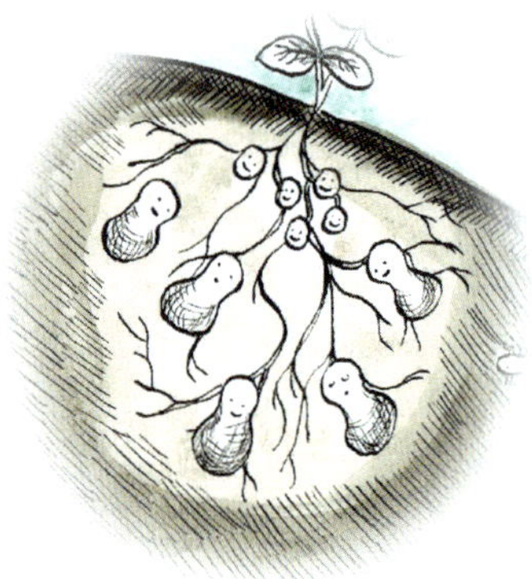

Peanut
The Fabulous
Fabacea!
Lentil

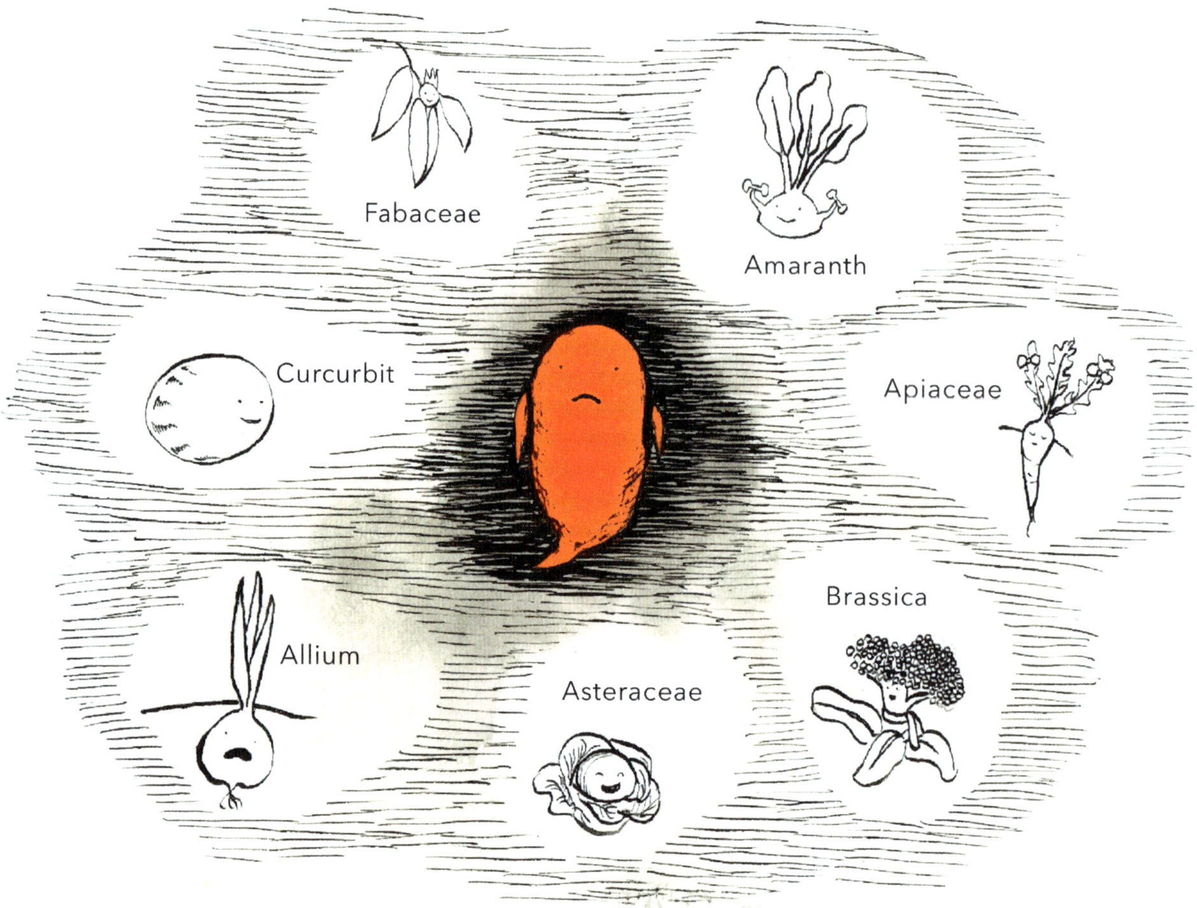
Fabaceae
Amaranth
Curcurbit
Apiaceae
Brassica
Allium
Asteraceae

By now Sweety was thoroughly disappointed. She had met so many interesting families on her journey, yet she didn't belong to any of them.

You should ask the **Morning Glory** family," whispered Mrs. Pea to Sweety. "I hear they like to get up early and watch the sunrise, just like you."

So Sweety walked and walked until sunrise, when she finally came across a family who was wide awake. She saw many plants; some had beautiful flowers, each with five petals, and some had winding stems and large, flat leaves. At last, underground, she saw some plants that looked just like her.

"Are you my family?" she asked.

Yes, Sweety," said Mrs. Sweet Potato giving Sweety a big hug. "We are so happy that you're finally here. We're the **Convolvulacea** (kon·vol·vu·la·SEE·uh) family, but you can call us the **Morning Glories** because we like to rise early every morning and open our flowers to greet the sun. Our family is large and consists of the beautiful flowers and vines you see around you. Sweet Potatoes are special plants in our family because we like to grow underground... just like you!"

Con-vol-vul-a-cea
I think I'll stick with Morning Glories.

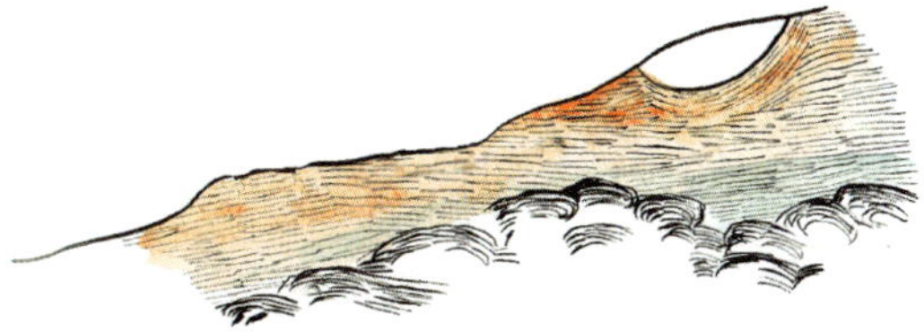

Sweety was so happy to finally meet her Sweet Potato family. They got up early to watch the sunrise together each morning.

They played cards in the afternoon and told stories over tea. Sweety was content, but she soon started to miss the Potatoes who she grew up with. She knew that she wanted to go home.

Sweety said goodbye to the Sweet Potatoes and the rest of the **Morning Glory** family, promising to visit them very soon. Then she started her long journey back home.

When she finally arrived at home, she was greeted by the entire **Solanaceae** family. The Potatoes, Tomatoes, Peppers and Eggplants took turns giving her hugs and asking questions. She shared stories of her great adventures until bedtime.

That evening, her parents read an extra long bedtime story and gave her a big kiss good-night.

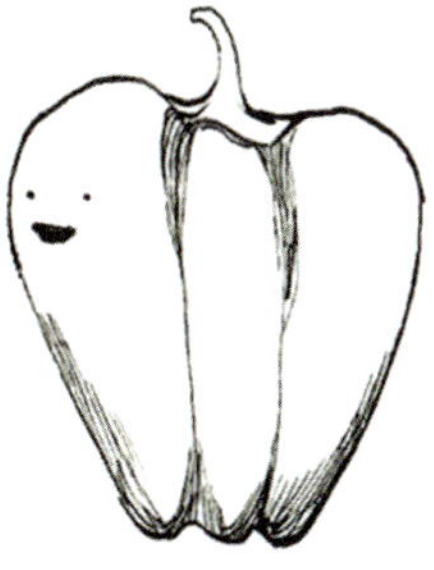

As Sweety drifted off to sleep, she realized that she had found her family and that she was finally home sweet home!

Katie's Favorites

The Wormfarm Institute: Programs intertwining art and agriculture
wormfarminstitute.org

Fermentation Fest: A rural art and farming festival
fermentationfest.com

The Land Stewardship Project: Supporting new farmers and conservation practices
landstewardshipproject.org/

FoodCorps: Teaching children about healthy eating in schools
foodcorps.org/

Good Food Jobs: Search for careers related to cooking, farming and organizing
goodfoodjobs.com/

MOSES Organic Farming Conference: The largest networking event for organic farmers in the U.S.
mosesorganic.org/conference.html

Visit www.Katie-Godfrey.com for more information on organic farming and her latest projects!

KATIE GODFREY's short stories have been featured in ***Greenhorns: 50 Dispatches from the New Farmers' Movement*** as well as other food and farming journals. She was fortunate to have the opportunity to work on organic farms in the Driftless region of Wisconsin, an area rich with creative people and beautiful landscapes. Katie was raised in Madison, Wisconsin and now lives in Albert Lea, Minnesota.

JULIA GOOTZEIT is an instructor at an early child education center in Washington, D.C. She illustrated ***Sweety Finds Her Family*** while working on a small farm in Reedsburg, Wisconsin. In her spare time she loves drawing, reading, and eating asparagus, and currently lives in Washington, DC with her partner and two rambunctious cats.

Katie and Julia met during an artist residency program run by the Wormfarm Institute, a nonprofit dedicated to the integration of culture and agriculture. The story and artwork were inspired by their daily farm chores: picking beans, growing sunflower seed sprouts and preserving tomatoes. Katie and Julia look forward to their next collaboration.

Interested in learning Spanish? Do you know a Spanish speaking child?

Sweety Finds Her Family is also available in neutral Spanish, translated by SARAI GUTIÉRREZ RODRÍGUEZ. Sarai is a translator certified by the Spanish Ministry of Foreign Affairs. In addition to her translation skills in the legal and financial field, Sarai is an expert on Literary Translation thanks to her degree obtained by the Universidad Pompeu Fabra. A huge lover of words and traveling, Sarai is also an English, Spanish and French teacher. She has lived and worked in Spain, France, and Great Britain. During her free time, Sarai likes to discover the world and its cultures, read, and walk by the beach. She currently lives in Fort Lauderdale, Florida.

30561371R00044

Made in the USA
Lexington, KY
07 March 2014